MEAT

Elizabeth Clark

Illustrations by John Yates

Carolrhoda Books, Inc./Minneapolis

All words that appear in **bold** are
explained in the glossary on page 30

First published in the U.S. in 1990 by
Carolrhoda Books, Inc.

Library of Congress Cataloging-in-Publication Data

Clark, Elizabeth.
Meat.

Includes index.
Summary: Describes where meat comes from, how animals
are raised to provide it, and how it is prepared for
food.
1. Meat—Juvenile literature. 2. Meat industry and
trade—Juvenile literature. [1. Meat. 2. Meat industry
and trade] I. Yates, John, ill. II. Title.
TX373.C53 1990 641.3′6 89-488
ISBN 9-87614-375-3 (lib. bdg.)

Printed in Italy by G. Canale C.S.p.A., Turin
Bound in the United States of America

1 2 3 4 5 6 7 8 9 10 99 98 97 96 95 94 93 92 91 90 89

Contents

What is meat?

Meat is eaten throughout the world. Many different animals are bred for meat, which is animal flesh used as food. Meat is mainly muscle—the tissues in the body used to produce movement. Some of the most popular meats are beef and veal (from cattle), pork and bacon (from pigs), poultry (from chickens and other birds), and lamb.

The amount of meat people eat varies greatly from country to country. In Western countries, a great deal of meat is eaten. But in some parts of the world, particularly the developing countries, very little meat is eaten. This may be for economic or religious reasons.

The production of meat is an important part of modern farming. The countries that produce the most meat are the United States, New Zealand, and Argentina.

rabbit

pig

goose

Meat in the past

It is believed that early humans lived on a diet of plants and small animals. At first they killed wild animals to defend themselves from attack, but when they found that animal flesh made good food, they became able hunters.

About 12,000 years ago, people began to keep animals so that they would have a steady supply of meat. But they had to keep moving from place

People have eaten meat for thousands of years. At this market in Nigeria, goats are being sold for their meat.

to place to find food for their animals. Soon, however, people began to live as farmers, settling in one place to grow crops and raise animals for food.

Gradually, people found that they could make their animals stronger and healthier through careful **breeding**. They mated their best male and female animals so that the best features of the parents would be passed on to the offspring. In later years, farmers experimented with **cross-breeding**, or mating different breeds. Many of today's popular breeds were produced through crossbreeding.

These two pictures show how one breed of sheep was improved by careful breeding in the eighteenth century.

Cattle, sheep, and pigs

Cows and bulls that are bred and kept on farms and ranches are called **cattle**. Cattle have been farmed for about 8,000 years, both for their meat and their milk. Some of the main beef breeds in the United States are the Hereford, the Aberdeen-Angus, and the Brahman.

The meat from calves (the young of cattle) is called **veal**. Veal calves are kept in small crates to limit their movement. This makes their meat pale and tender. Many people believe the way calves are raised and killed for veal is cruel.

Sheep have been farmed for their meat, wool, and milk for about 12,000 years. Sheep breeds raised for meat include the Suffolk, the Shropshire, and the Southdown.

Pig farming dates back about 9,000 years. Pork, ham, bacon, and many types of sausages come from pigs. Some popular American pig breeds raised for their meat are the Hampshire, the Duroc, and the Yorkshire.

A Texas Longhorn bull, a Southdown sheep, and a Yorkshire pig

Poultry, game, and other animals

Poultry are birds raised for meat or egg production. This includes chickens, turkeys, geese, and ducks. Chickens, which were first farmed about 2,400 years ago, are eaten in many parts of the world. Turkeys are popular during the Christmas season in many Western countries and for Thanksgiving in the United States.

Game refers to all wild birds and animals

hunted for pleasure or for their meat. Game birds include pheasants, partridges, and quail. Deer are game animals that are often hunted for pleasure, and some people also enjoy deer meat, which is called **venison**.

Other animals, such as rabbits and hares, are also raised for their meat. In Scandinavia, reindeer are eaten, while in France, many people eat horse meat. Goats, which are cheaper to raise than cattle, are eaten in some of the developing countries of the world.

Meat as food

This diagram shows the different amounts of nutrients found in meat.

Meat contains many nutrients that our bodies need to stay healthy, including protein, fats, vitamins, and minerals.

Meat is sometimes described as either red or white. Red meat is meat that is dark in color, such as beef and lamb. White meat, such as chicken and turkey, is lighter in color.

Red meat is high in saturated fats. Eating too

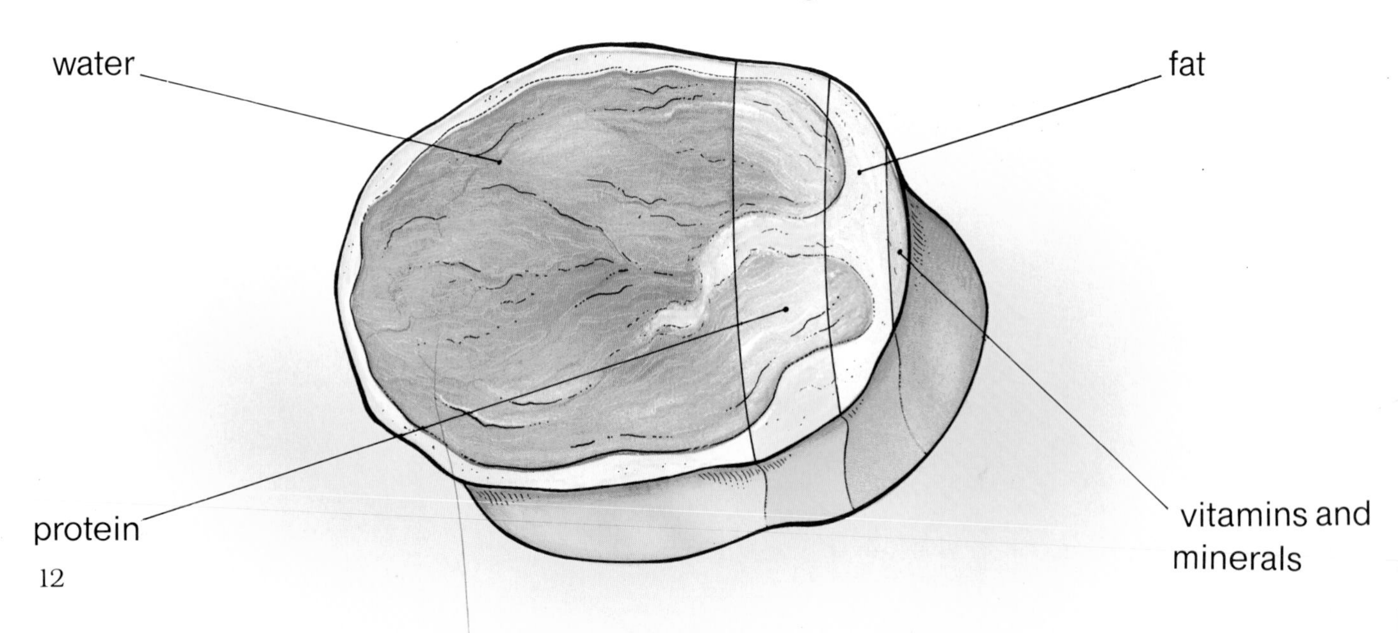

Right: Chickens on sale at a market in Mexico
Below: "Fast foods," such as hot dogs and hamburgers, are high in fat and therefore not good for us.

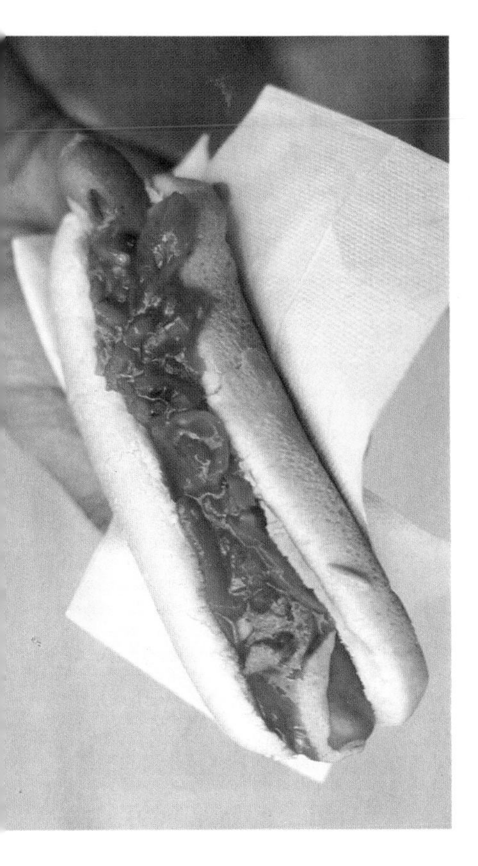

much of these fats can clog the body's blood vessels and cause heart disease. In Western countries, where people eat a lot of red meat, heart disease is common. In countries where people eat mainly fish and plant products, fewer people develop heart disease.

Chicken and turkey are lower in fat than red meat. Health experts say we should eat more white meat and cut down on red meat. Also, fat should be trimmed from red meat to make it healthier.

The liver, heart, kidneys, brains, and other organs of some animals are edible. These organs, known as **variety meats**, are a good source of nutrients but are very high in fat.

Raising animals for meat

Animals raised to provide meat are bred to produce high-quality meat and to gain weight very quickly. When the animals have reached a certain weight, they are said to be **finished**. This means they are ready to be slaughtered and sold. Cattle are finished when they are a year old, and lambs when they are between 3 and 6 months old. Pigs bred for pork are finished at about 16 weeks, while pigs bred for bacon are usually older. Chickens are finished when they are 6 weeks old.

Above: Intensive chicken farming in Nigeria

Right: An Australian sheep farm where the sheep are raised for their meat

Beef cattle are herded into a feed lot on this American cattle ranch.

Many of today's pig and chicken farmers use a method called **intensive farming**. Hundreds of pigs or chickens are crowded into specially designed buildings. They are fed frequent, high calorie meals and have no room to exercise, so they gain weight quickly. Many people believe that it is cruel to keep animals in such cramped living conditions, and some farmers now keep their animals in open spaces, allowing them to move around freely. This is known as **free-range farming**.

From farm to store

When animals are finished, they are sometimes taken to a farmer's market to be sold. Usually, however, they are taken directly to the slaughterhouse, where they are killed quickly and with as little suffering as possible.

In many countries, after the animals are slaughtered, they are examined to make sure that they are free of disease. Then some meat

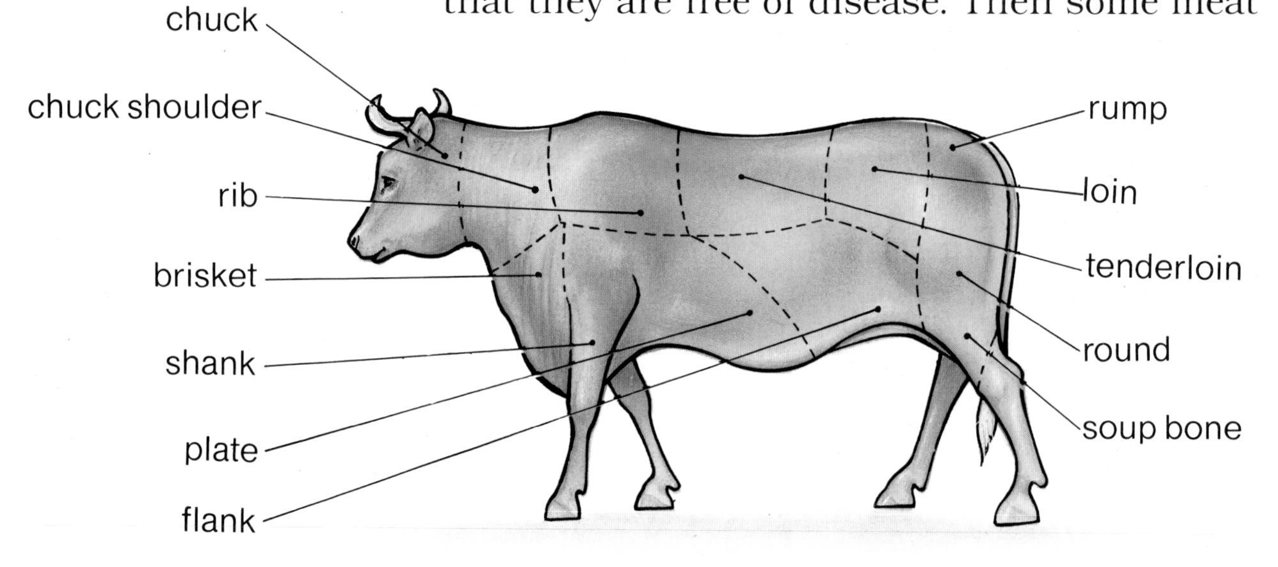

is sent to factories to be made into sausage, bacon, and other products, while other meat is refrigerated and sent to butchers. Butchers cut meat into parts and sell it.

These diagrams show which part of an animal's body the different cuts of meat come from.

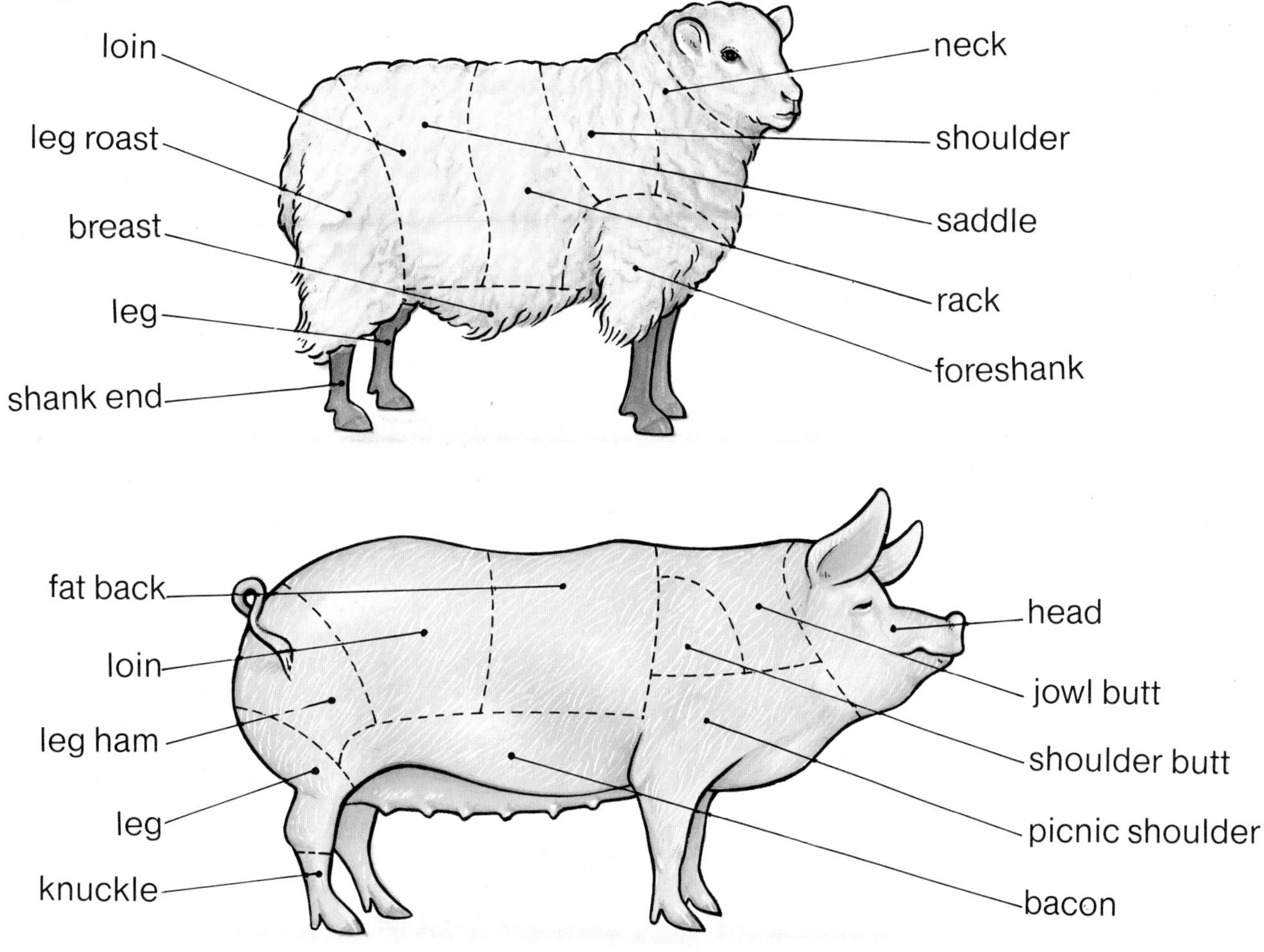

Keeping meat fresh

Meat must be eaten when it is fresh, or tiny organisms called bacteria will multiply in it, causing the meat to decay. Eating meat that has decayed is unhealthy and can be fatal.

Refrigeration allows us to keep meat fresh for long periods of time. Meat may be kept chilled in refrigerators or deep frozen and stored in

Meat can be preserved by any of these different methods.

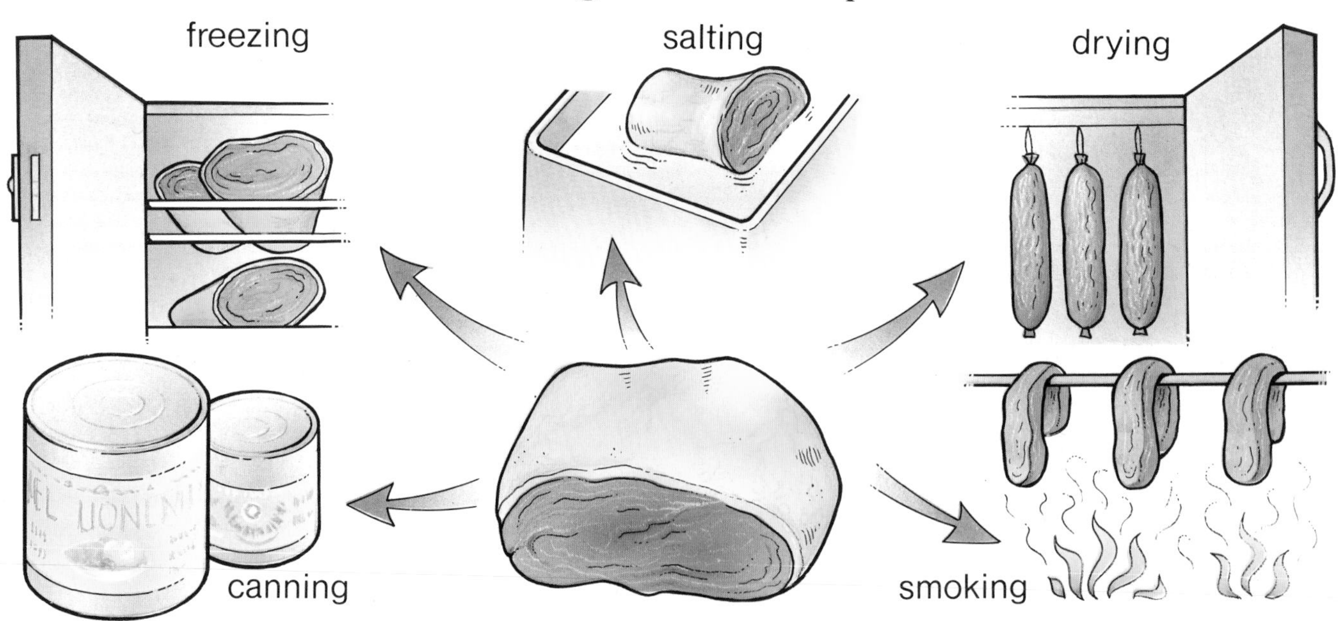

Right: Preserving ham in salt in Andalusia, Spain

Below: Meat hung up to dry in the wind in Afghanistan

freezers. Butchers have giant refrigerators for storing meat.

Meat can also be preserved by canning. The meat is sealed in an airtight can and heated. The heat destroys bacteria already present in the meat, and the airtight seal keeps more bacteria from entering the can.

Before refrigeration and canning were invented, people preserved meat by salting and drying it. These methods are still used today. Bacon and ham, for example, are **cured** by salting and drying (and also sometimes by smoking over a fire). All of these methods ensure that the meat is safe to eat.

Barbecues are very popular in many countries. These men are grilling sausages in a park in Australia.

Cooking meat

Meat is almost always cooked before it is eaten. Cooking kills any bacteria that remain in raw meat and makes meat easier to digest. Also, most people prefer the taste of cooked meat. Here are some common methods of cooking meat.

Meat can be roasted by placing it in a dish and cooking it in the oven. Fried meat is cooked on the stove in a pan of hot oil or in its own fat. In some Asian countries, meat is stir-fried by stirring it for a few minutes in a special pan called a wok, in a small amount of very hot oil.

Cooking meat slowly in liquids flavored with vegetables and herbs is called stewing. Meat curries are a type of spicy stew that is eaten in India and other Asian countries. Meat can also be cooked on a grill over an open flame.

Barbecuing meat, or cooking it outdoors on a grill, is very popular in the United States and in other countries, such as Australia. Another way to cook meat is to broil it in a pan in the broiler section of the oven.

Above: A roast lamb dinner in the Middle East

Left: Cooking kebabs at a carnival stand

Beliefs about meat

Many people throughout the world eat no meat at all; these people are called **vegetarians**. Many people today choose to be vegetarians because they believe it is wrong to kill animals for their meat. Others believe it is healthier not to eat meat.

Vegetarians live on a diet of plant and dairy

Hindus believe that cattle are sacred animals. This cow is specially decorated for a Hindu ceremony.

A butcher sells meat that has been prepared according to Muslim religious laws.

products, though some vegetarians, called **vegans**, eat only plant products. Experiments to find meat substitutes, such as soybeans, have been carried out for many years.

Some religions forbid eating the meat of certain animals. For example, Hindus believe that the cow is a sacred animal, and they do not kill it for meat. Some Hindus and Buddhists do not eat any meat at all because they believe that it is wrong to kill living creatures.

Muslims and Jews are forbidden by religious laws to eat the meat of pigs or of animals and birds that eat meat. They may eat the flesh of cloven-hooved animals such as cattle, sheep, and goats, which eat grass and plants. The meat must be prepared in a special way. Meat

A colorful selection of meats you might see in a butcher's shop. Some experts believe that in the future our diets may have to contain less meat.

Soybeans can be used as a meat substitute. These children are sorting soybeans from the harvest in Thailand.

prepared according to these religious laws is called *halal* by Muslims, and *kosher* by Jews.

Meat farming is an expensive way of using the land because animals raised for their meat need so much land for grazing. Some experts believe that in the future we may have to adapt to a diet containing more plant foods and less meat, so that more land can be used to grow crops to feed people.

Make sure there is an adult nearby when you are cooking.

Hamburgers:

You will need, for 4 burgers:

1 lb. ground beef
salt and pepper
4 hamburger buns
sliced onions (optional)
sliced tomatoes (optional)

1. Put the meat in a mixing bowl. Blend in a pinch of salt and pepper.

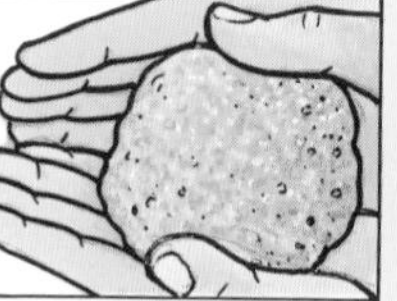

2. Put the meat on a clean surface. Divide it into four equal portions. Shape and flatten each portion into a patty about ½ inch thick.

3. Put the meat in a frying pan over medium heat. Fry for 3-4 minutes. Turn the patties over. Fry for another 3-4 minutes.

4. Place each cooked hamburger on a bun. Add the onions and tomatoes if you wish. You may also want to add ketchup or mustard.

Turkish lamb kebabs

You will need, for 4 kebabs:

1 lb. lamb diced into 1 inch cubes (you can buy it already diced or ask the butcher to dice it for you)
8 small button mushrooms
8 small shallots
8 cherry tomatoes
1 red pepper, seeded and chopped
melted butter or cooking oil in which some rosemary has been soaked
Equipment: four long metal skewers

1. Very carefully thread a cube of lamb onto a skewer. Then thread on a shallot, a mushroom, a piece of red pepper, and a tomato. Follow this with a piece of lamb and continue until the skewer holds 8-10 ingredients. Do the same thing with the remaining skewers.

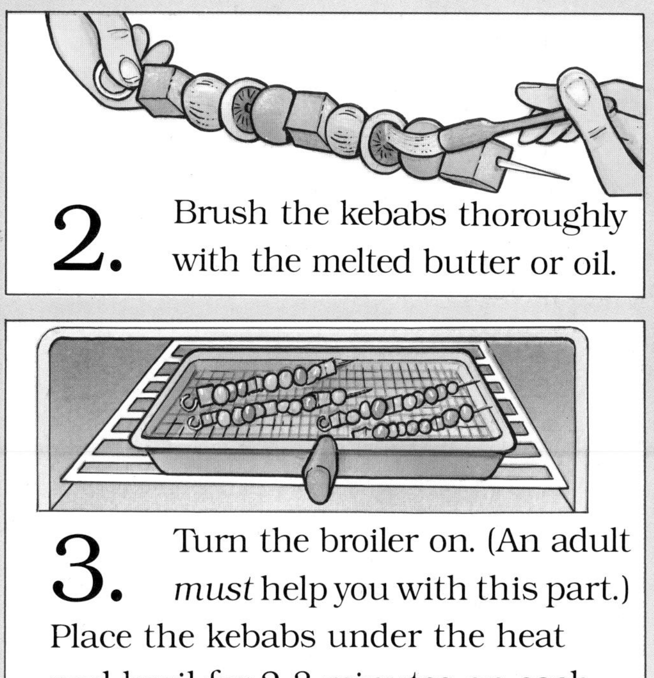

2. Brush the kebabs thoroughly with the melted butter or oil.

3. Turn the broiler on. (An adult *must* help you with this part.) Place the kebabs under the heat and broil for 2-3 minutes on each side. Lower the heat and cook for another 8 minutes, turning the skewers occasionally.

4. Remove kebabs from the broiler and serve.

Baked beans with bacon

You will need, for 3-4 people:

½ lb. bacon with rind and fat trimmed off
½ cup sliced mushrooms
1 medium onion, chopped
1 small red pepper, seeded and chopped
8 oz. can of baked beans
salt and pepper
1 teaspoon chili powder
3 tablespoons cooking oil

1. Preheat the oven to 350°F.

2. Cut each slice of bacon into thirds.

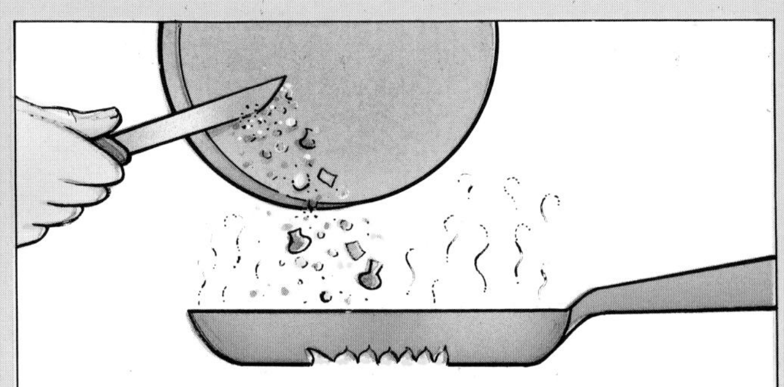

3. Heat the oil in a frying pan and add the bacon, mushrooms, onion, and red pepper. Fry until the bacon looks brown. Add salt, pepper, and chili powder.

4. Put the bacon mixture into a casserole or baking dish, pour on the baked beans, and cover. Bake for 30 minutes.

5. Using oven mitts, take the casserole out of the oven and serve. (Sliced boiled potatoes go very well with this dish if you want to make it more filling.)

Glossary

breeding: mating animals to produce the best quality offspring

cattle: cows, bulls, and calves that are raised on farms or ranches

crossbreeding: mating two different breeds to obtain the best features of both

cured: preserved by salting and drying

finished: ready to be slaughtered

free-range farming: farming animals in open spaces and natural conditions

game: wild birds and animals that are hunted for pleasure or for their meat

intensive farming: farming animals in specially controlled conditions

poultry: chickens, turkeys, and other birds raised for meat or egg production

variety meats: livers, kidneys, and other animal organs that are eaten

veal: the meat of calves

vegans: vegetarians who eat no animal products at all

vegetarians: people who eat no meat

venison: deer meat

Index

Photo acknowledgments

The photographs in this book were provided by: pp. 6, 19 (top), 20, 21, J. Allan Cash; pp. 7, 13, 20, 24, Wayland Picture Library; pp. 14, 15, 22, 23, 25, Hutchison Library; pp. 19 (bottom), 21 (top), Christine Osborne.